Discover a... SENSES AT THE PARK

Ursula Pang

We have five senses. They are seeing, hearing, feeling, smelling, and tasting.

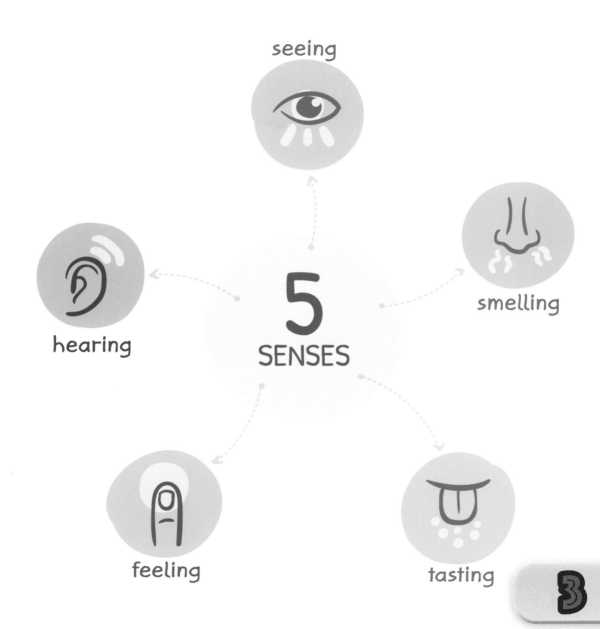

I see colorful balloons at the park.

I see dogs playing at the park.

I hear music at the park.

I hear people laughing at the park.

The sun feels warm at the park.

The grass feels soft at the park.

I smell flowers at the park.

I smell popcorn at the park.

I taste ice cream at the park.

Ice cream feels cold!
It tastes sweet!

Published in 2023 by The Rosen Publishing Group, Inc.
2544 Clinton Street, Buffalo, NY 14224

Copyright © 2023 by The Rosen Publishing Group, Inc.

All rights reserved. No part of this book may be reproduced in any form without permission in writing from the publisher, except by a reviewer.

First Edition

Editor: Greg Roza
Book Design: Michael Flynn

Photo Credits: Cover photoiva/Shutterstock.com; p. 3 Jimena Catalina Gayo/Shutterstock.com; p. 5 Svetlana Satsiuk/Shutterstock.com; p. 7 otsphoto/Shutterstock.com; p. 9 IRINA ORLOVA/Shutterstock.com; p. 11 SeventyFour/Shutterstock.com; p. 13 Serg64/Shutterstock.com; p. 15 muay26/Shutterstock.com; p. 17 Igor Normann/Shutterstock.com; p. 19 krsmanovic/Shutterstock.com; p. 21 Tatyana Vyc/Shutterstock.com; p. 23 Wirestock Creators/Shutterstock.com.

Library of Congress Cataloging-in-Publication Data

Names: Pang, Ursula, author.
Title: Senses at the park / Ursula Pang.
Description: Buffalo, New York : PowerKids Press, [2023] | Series: Discover at the park!
Identifiers: LCCN 2022026191 (print) | LCCN 2022026192 (ebook) | ISBN 9781538388952 (library binding) | ISBN 9781538388938 (paperback) | ISBN 9781538388969 (ebook)
Subjects: LCSH: Senses and sensation--Juvenile literature. | Parks--Juvenile literature.
Classification: LCC QP434 .P36 2023 (print) | LCC QP434 (ebook) | DDC 612.8--dc23/eng/20220606
LC record available at https://lccn.loc.gov/2022026191
LC ebook record available at https://lccn.loc.gov/2022026192

Manufactured in the United States of America

Some of the images in this book illustrate individuals who are models. The depictions do not imply actual situations or events.

CPSIA Compliance Information: Batch #CWPK23. For further information contact Rosen Publishing at 1-800-237-9932.